PISCICULTURE.

AN ADDRESS

ON THE

Artificial Breeding of Fish,

THEIR

HABITS, ETC.

DELIVERED BEFORE THE

DETROIT SCIENTIFIC ASSOCIATION,

BY

N. W. CLARK,

OF NORTHVILLE, MICH.

DETROIT:
TRIBUNE PRINTING COMPANY,
WILLIAM A. SCRIPPS.
1875.

PISCICULTURE.

AN ADDRESS

ON THE

Artificial Breeding of Fish,

THEIR HABITS, Etc.,

DELIVERED BEFORE THE DETROIT SCIENTIFIC ASSOCIATION,

BY

N. W. CLARK, of NORTHVILLE, MICH.

In this essay I cannot hope, with my limited knowledge of natural history, to add much to the scientific lore of our country, but I wish to contribute my mite of practical experience in the direction of investigations now being prosecuted in our country relating to the propagation, and especially to the sustenance of fish during the first few months of their life.

By pisciculture, in the ordinary use of the term, is meant the artificial propagation or raising of fish with a view to an increase of the supply for food. The waters, as well as the soil of the earth, are largely drawn upon for the subsistance of mankind. The savage visits the streams and the lakes no less than the woods for the game upon which he lives. And man in a civilized condition of life depends nearly as much upon the products of the fisheries as upon those of the fields.

Though the land in a state of nature yields each of the various kinds of fruit and grains employed as food, still the supply is only sufficient to maintain man and ănimals living purely in a state of nature. The principle is simply that of the equilibrium of demand and supply as a natural condition of things. Whenever by the various vicissitudes of season, of soil and of climate, the demand exceeds the natural supply, the equilibrium is preserved through the destruction by famine and by violence of a portion of the population.

This is the so-called struggle for life, which is more or less observable everywhere among men and animals in a state of nature. Thus the demand is regulated by the supply.

In other words supply is the independent variable, and demand the dependent one.

Under a civilized condition of life, on the other hand, supply is regulated by demand, rather than demand by supply.

The wants of mankind, instead of the means of supplying want, constitute the independent quantity. The supply is constantly being compared with the demand as a standard.

If supply is found to be below necessary demands the former must needs be increased since the latter cannot be diminished.

It is thus that the pursuit of agriculture has arisen. It is simply to increase the natural product of the land and thus bring supply up to the level of the demands of civilized life among an ever increasing population, that we have the various arts and appliances, the industry and providence of successful farming. .Similarly, the object had in view in pisciculture is to augment the supply of another and important kind of food by increasing the natural product of the waters.

This branch of industry has received, until recently, comparatively little attention.

Agriculture has naturally been developed long antecedently to pisciculture. Man's natural home is upon the land. His attention would thus be earlier directed to the culture of the soil than to that of the water. The principles underlying the successful raising of crops and flocks naturally become familiar to him before those relating to the production of fish.

Indeed, it was not until it it began to be realized that the supply of the finny product must soon become sensibly diminished, that we began, in this country, to direct our attention to the matter of preserving and increasing the product of the rivers and lakes.

It is thus that we have laws to prevent the drawing of the seine in our inland lakes.

It is thus also that our State appropriates money

to the support of the artificial hatching and distribution of some of the most important and valuable kinds of fish. In older countries, as in Europe, attention has been now for many years directed to the systematic raising of fish for the market. In Italy and Egypt the art of pisciculture dates from a most remote antiquity.

With the Chinese, owing to the dense population, and the extended water territory of China, the raising of fish forms a very common and important branch of industry. Pisciculture, as now carried on, originated with the French, being introduced by M Remy, a fisherman, whose means of support was fishing in the streams of La Bresse, in the Vosges.

It was the great waste of eggs observed to be attendant upon the natural breeding of fish that led M Remy to experiment upon an artificial method of treating the ova, with a view to restocking the streams of his native district.

His success at once attracted the attention of many of the leading men of the country; prizes and preferment were the fisherman's, and his enterprise became the object of governmental patronage. Thus in a period of little over twenty years artificial fish culture has so developed in France that an immense establishment has been put in operation at Huningue, on the Rhine, at which millions of eggs are annually received, and from which, after hatching, the young fish are distributed to Germany, Spain, England and other countries throughout the Continent.

The total number of all kinds of fish sent out

from Huningue during the first ten years was upwards of 110 millions.

At other points also the artificial system is extensively employed. The salmon fisheries of the river Tay and those of Ireland, those of the Thames and of Scotland are notable examples.

The number and extent of establishments devoted to the propagation of the various kinds of fresh water fish are rapidly increasing in the old country, and the importance of systematic pisciculture is becoming more and more a subject of governmental consideration. Such will be the case in our own country also, as it becomes older.

Already there are several more or less extensive fisheries in successful operation in different parts of our own country.

The Legislature of our own State, two years ago, appropriated $7,500 per annum to inaugurate the enterprise of fish culture in this State.

The Governor, in his recent message, says: "The success attending this work, thus far, is sufficient proof that fish culture in Michigan is no longer an experiment." He advises the continuance of the appropriation, saying: "There can be no doubt that in a very few years the returns from the expenditure will prove it a wise and prudent investment."

In this State the direction of fish culture is very naturally toward the propagation of the white fish of our great lakes. Indeed, it was owing mainly to a proper estimate of the importance and value to the State of the whitefish fisheries upon those lakes, coupled with the fact that the supply of this kind of fish was becoming sensibly diminished and must before long altogether fail, that attention in Michigan began to be called specially to the subject of fish culture.

As in other departments of labor, particularly in those engagements which involve considerations relating to the natural order of things, the successful prosecution of pisciculture requires a careful induction of conditioning principles together with much experience in the application of those principles to different cases, both under the same and also under diverse circumstances. We can become acquainted with the principles underlying successful pisciculture only by observing the conditions and circumstances attending the inception and development of fish in their native haunts.

Much wasteful expense and discouraging failures have resulted from disregard of this simple and obvious fact. Nature is the only proper teacher of her own laws, and true methods can be learned in no other school of art than that in which she presides. As the farmer needs to know the nature of the soil on which a given plant spontaneously grows, the circumstances of climate, season, etc., under which it naturally develops; the nature, habits, etc., of the plant itself, so the pisciculturist must understand the peculiarities of the water which a given kind of fish naturally inhabits; whether it is salt or fresh, pond or spring, lake or brook, etc., must know the time of year, temperature of water, etc., in which the parent fish spawn and in which the ova hatches; must know also the period of the self-sustenance of the infantile hatching, its natural food after the expiration of this period, its nature and habits throughout its entire development to full maturity.

A practical illustration of the importance of these considerations is furnished in the raising of each and every kind of fish known to pisciculture. An example in point is cited from the writer's own experience in the treatment of white-fish, as published in the *Forest and Stream*, over his own signature, bearing date, Clarkston, June 29, 1874, as follows: "About November 15th, 1869, through the kind assistance of Mr. Seth Green I placed 50,000 white-fish ova in my hatching troughs, in spring-water, at a temperature of 47° Fahrenheit, at this place where I was then hatching brook trout successfully. I soon discovered that these eggs were entirely different in their nature from those of the brook trout, for within two weeks' time, nine-tenths of them had turned white and were worthless, and I was about to abandon them in despair.

At this critical time, Mr. Green came to my relief, and after a careful investigation, we found a small portion had indications of vitality in them and he advised me to carefully remove all the good eggs, and by this means I succeeded in hatching about 2,000 in good condition Jan. 15th.

These fish were entirely different from the young trout when first hatched out, as the moment they emerged from their shell they darted off and exhibited a rapid motion in the water, while the latter were quite inactive, owing to the fact that young trout retain a large umbilical sac, which sustains them some sixty days without food, while the white fish have a very small appendage, which is absorbed in about fifteen days.

These escaped from my charge through the meshes of No. 12 wire cloth, down the stream, and I had no opportunity to experiment further with them that year. On the following November, 1870, Mr. George Clark, who is an intelligent and experienced fisherman, kindly aided me in securing the same number of the ova from the Detroit River that I took the previous year.

I placed these in the same water as before, and succeeded in hatching a much larger proportion, and from my previous experience I selected No. 40 copper wire cloth, which proved effective in retaining them in their troughs.

This gave me an opportunity to use all the skill possible to keep them till spring.

Soon after these were hatched, James W. Milner, U. S. Deputy Fish Commissioner, visited my hatchery, and we decided that he should take home with him 100 of these swift-motioned fellows, and the balance, being some 3,000, were to remain in their hatching boxes.

Our plan was to learn what artificial food would best sustain them till spring.

As Mr. Green had not yet learned what they required at this infantile period, all our efforts failed, as all died within four weeks, notwithstanding our constant watchfulness over them. This result quite puzzled me, and I began to study the causes that produced this failure, and as I knew that the water in which the parent fish naturally deposit their ova about the shoals of our great lakes, becomes frozen over about the middle of November, and remains so until about April 1st, it occurred to me that the low temperature of the water in which these eggs laid from the middle of November till April 1st (being at a uniform temperature of from 32½ degrees to 33 degrees), retarded the process of their incubation to the season of the year when the ice leaves the shoals and the animalculæ develops sufficiently to sustain them, which is about the time their umbilical sac, being absorbed naturally, disappears.

These ideas which suggested themselves to my mind, led me to try practically and prove the truth of my theory; consequently I caused to be erected a large hatching house, in the Fall of 1871, and took water from a pond raised on a small stream which became frozen over early in November and remained so till April, at which time they hatched out.

The water that flowed over these eggs during that time, stood at a uniform temperature of about 33 degrees. A much larger proportion of these ova hatched out than previously, and remained vigorous and healthy till the time they were planted in some of the desirable lakes of this county (Oakland). A good number were also placed in the Detroit river.

This natural and scientific method settled the question in my mind that I had discovered the only true mode that would result in perfect success. In this connection, as bearing upon the question of the natural food for the young fish, immediately after the absorption of the umbilical sac. I give the following minutes furnished me by Dr. P. N. Hagle, an experienced microscopist associated with me in the experiments.

For five years I have directed my attention particularly toward the discovery of artificial food, or otherwise, the natural aliment of the white fish during its infantile stage, and my experiments and observations have been attended with results exceedingly satisfactory to myself, and as I feel, of the greatest importance in their bearing upon the interests of successful pisciculture everywhere. I will attempt to give you, briefly, a description of the method I have pursued.

About the middle of November, 1873, I placed some 1,800,000 of the ova of white fish in my hatching boxes, which received a constant supply of running water at a temperature of 33 degrees.

I also arranged in the same building, an additional trough, through which flowed a stream of spring water at an uniform temperature of 46 degrees.

Knowing that the eggs would develop and hatch in water of the latter temperature in sixty days, while in that of the former it required one hundred and thirty-five days, I was led to adopt such a mixing of the two waters as should graduate the hatching to such times as I desired.

I accordingly placed about 3,000 ova in the spring water and hatched them about January 15th. Of these I placed all except some 200 in a small lake near my residence. The 200 were placed in a tank of running spring water in which they survived about four weeks.

Again there hatched on February 15th about 6,000 more, which were placed in part in the lake and a portion in the tank, as I had done before. The latter in this case lived about five weeks, having, as I suppose, better food than those previously placed in the tank, yet not sufficient to sustain them, so that they also literally starved to death.

I continued, however, still further, hatching the balance of the 1,800,000 about April 1st. Of these I put, April 9th, 200 into a tank receiving a constant supply of lake water, and the same number into a tank of spring water, having taken pains in each case to wash and remove from the tank all the slime in which might be retained any remains of animalculæ or insect life.

Hitherto the young fish had began to die at the end of three weeks and at the expiration of four weeks were all dead.

The theory which suggested the observations to which I am now about to refer, has been already hinted at in the extract from the article published as above quoted, in which was maintained that the fish should be hatched at such a time of year that the umbilical sac on disappearing should be immediately followed by the animalculæ as a proper and natural food of the young fish, and which could not occur until the opening of the warn season.

These considerations led me to make the examinations noted in the following memorandum.

DATE.	SPRING WATER.	LAKE WATER.	REMARKS.
1874. May 8.	Temperature not noted.	Temp. 52° Fah. No animalculæ to be seen.	Germs of animalculæ, however, were detected.
May 10.	Temp. 41°. No animalculæ to be seen.	Temp. 60°. Abundance of animalculæ.	The lake water taken from a still point, near the shore, also from the tank.
May 14.	Do.	Temp. 60°. Animalculæ increased in size and numbers.	Temperature noted at the fount, and found to be a degree lower than at the tank. Time, 7:45 A. M.
May 15.	Do.	Temp. 53°. Nothing seen.	Vision very indistinct by reason of clouds. Night preceding was cold. Time, 8:15 A. M.
May 18.	Temp. 44°. No. animalculæ visible.	Temp. 53°. Great activity of animalculæ.	The fish in the lake water in much better condition than those in the spring and have increased to double their size when hatched. Observations made at 5:15 P. M.; good light.
May 22.	Temp. 41°. No animalculæ	Temp. 53°. Other forms of animalculæ have made their appearance in great numbers.	Time, 8:30 A. M.
May 26	Temp. 42°. Nothing visible.	Temp. 53°. Great numbers of animalculæ seen.	Many fish have died in the spring, while all seem to be doing well in the pond, having increased to three or four times their size at hatching.
May 28.	Temp. 44°. Nothing visible.	Temp. 60°. Numerous varieties of animalculæ.	Fish still doing well in the lake water. A fine clear view at 4:15 A. M.
June 2.	Temp. 56° at 100 feet from the spring. Where the water enters the tank nothing visible.	Temp. 56°. Animalculæ lively and of a minute radiate appearance.	Fish are twenty times their original size. Have had heavy showers which have raised the temperature of the spring.
June 5.	No observation.	Temp. 60°.	Light bad. All the fish in the lake water doing well. Some in spring water found dead from destitution of food, as evinced by their transparency under the microscope.
June 8.	Temp. 54°. Animalculæ visible.	Animalculæ visible and very lively.	The fish may be seen gathering about the point of entrance of fresh water, coming to them through a perforated tube. They seem actively engaged in pursuit of food, which might seem to be more abundant at that point.
June 18.	Temp. 48°. Animalculæ still seen	Animalculæ lively.	The fish doing well. Those in spring water beginning to improve. Some of those in lake water 1½ inches in length Time, 8:30 A. M. Weather cloudy with rain.

These experiments have been conducted with great care, and seem to form the basis of an irrefragable argument that the young of any kind of fish cannot survive the vicissitudes of climate if means are used to hasten their incubation, so that they shall hatch out more than about three weeks before the balmy breezes from the south cause the animalculæ, which constitute the natural food of the young fish, to accumulate in the waters in sufficient quantity and proper quality for their sustenance.

The development of the animalculæ, their existence and growth, imperceptible though they be to the naked eye, are no less a product of the light and warmth of the spring sun, than is that of the vegetation—the grass that comes upon our fields for the sustenance of our flocks.

With these considerations in view, is it not obviously as true that other varieties of the salmonoid family, indeed all fish as well as the white fish, must meet the same fate of starvation and death where means are used to develop them into life out of their natural season, or at a time when the waters of this latitude are so ice-bound that not a particle of animal life can exist for their sustenance.

I wish to inquire of the practical pisciculturist, what reason he has to suppose that salmon ova, taken from the McCloud river in California, the temperature of which is reported to be 50° Fah., and their eggs being subject to the vicisitudes of climate during their transportation, after arrival here, which is about the middle of October—and the young fish thus become dependent upon such food as chance may furnish them, I wish to ask who can expect a single one of the fish hatched under such circumstances to survive in our latitude during the winter?

I believe it is just as necessary to the existence of this fish, as of the white fish to be hatched at that time of year, when the temperature of the water is such as to admit of the development of animalculæ the only natural food of the young fish after the absorption of the umbilical sac.

Take another case — the salmon trout for instance, which is a native of our state, and let us examine their natures and necessities.

About Oct. 1st, 1874, I visited Mackinaw for the purpose of procuring some of the ova of this fish for breeding purposes, and I suceeded in securing some 300,000.

I arrived at Northville with them about the 25th of the same month, the eggs being in good condition save a few thousand that were injured in transportation in the extreme warm weather. The eggs were placed in pure spring water at an average temperature of 45°. It is now Jan. 20th, and they are all hatched out, and when I look upon the beautiful lively little fellows, in full health and vigor, each with its exact quota which nature has furnished them appended to their stomaches, only sufficient, however, to sustain them for some fifty or sixty days, it makes me feel sad when I think of the destruction there must certainly be from famine if turned out in our barren waters at this season of the year, for want of that natural food which I believe is just as necessary to insure their proper and healthy growth as milk is for babes. All fish-breeders full well know that with all the care and the best artificial food we can furnish, only a limited number will survive this critical period, simply because they are found to eat, if anything, food unsuited to their age and condition. Let us now inquire, for a moment, what are the natural conditions of these fish in their native haunts. We find them spawning during the month of October, upon the rocky and gravelly shoals of the waters they inhabit. In the northern latitudes, where this kind of fish most abounds, we find, within a few weeks after the spawning season, the water frozen over, and thus the eggs lie in ice-water, which retards their hatching until about March 1st, when they come out, having somewhat the appearance of a tad-pole. In this condition they lie among the rocks and pebbles, well protected from the depredations of their enemies, until about the first of May, when their sac having been absorbed, the waters are sufficiently warm for the appearance of the animalculæ upon which the young fish are now become dependent. Thus, from natural considerations, from the laws regulating an universal hygiene, and also from the confirmation that the view has, in the testimony of sailors, who report that few dead fish are even found in the waters, we judge that nature's method is the only true one, and that it must be observed and imitated in order to attain any real success in any of the artificial processes of fish culture. Allow me here to relate a bit of my own experience in trout-breeding, as I feel confident you want facts. In the year 1872, being then fully engaged in experiments connected with the breeding of white fish, and only having saved a limited supply of the ova of brook trout, I arranged the few I had (about 2,500) in hatching-boxes in such a manner that when hatched out they could take possession of an adjacent pond, 14 x 45 feet in size and two feet deep.

These eggs received but little attention during the winter; they were late in hatching; and about April 1st, I observed for the first time numbers of them swimming about in a lively condition, their yolk sacs having almost entirely disappeared.

These active little fellows, about 2,000 in all, were allowed to remain there entirely unfed and uncared for until about September 15th, at which time they had attained a size far beyond any I had ever seen. I never saw a dead one among them. It will be borne in mind that this pond was spring-water, about one hundred feet from the spring and of a temperature of about 55°.

This was the same water in which I had experimented three or four years previously, in raising brook trout by following Seth Green's book on "Fish culture," feeding the young fish on lobered milk, chopped liver, etc. These fish I did not succeed nearly as well in raising as I did those above referred to, and I am satisfied that the former had but little if anything, to eat, but the animalculæ contained in the water. Now is it reasonable to suppose that had these fish been hatched out by December 1st as is sometimes done, they would have survived longer than the time of their self-sustenance or the period of the absorption of the umbilical sac—a term of about sixty days at most. I will relate another incident in my experience, which is of very great value as an aid in arriving at correct conclusions as to the conditions of growth and well being of fish. About October 10th, 1873, through the aid of Professor Baird, U. S. Fish Commissioner, there were sent to our State Fish Commissioners some 30,000 of the salmon ova from California. These were placed in charge of Jackson Crouch, near Jackson, Michigan, who was then breeding brook trout quite successfully.

These were hatched in apparent good condition about November 1st following, and some time early in January they were all placed in the pure streams of that vicinity, except 2,000 which were taken by Commissioner Jerome and placed in the spring at the State Hatchery, at Pokagon, the State Hatchery being at this time already erected at that place. Now for the result. I have recently visited and seen both lots, and find on close inspection and inquiry that those which have been under the care of Mr. Jerome have measured from six to seven inches in length, while those left with Mr. Crouch are only from four to six inches long and not in as good condition. The growth and condition of these fish can only be accounted for on the ground of difference in the quantity of food.

Mr. Crouch having, in addition to his 1,000 salmon, to procure food for 20,000 more ravenous brook trout, while Mr. Jerome had only the 1,000 to feed, and that at the expense of the State. The old adage holds true in fish culture as in agriculture, that good feed makes fat fish as well as fat calves.

How about the young salmon that were turned out to perish, as I fear they have, in the barren waters at this cold season of the year?

I was informed by Mr. Crouch that so far as he had any knowledge, none which he had turned out to shirk for themselves, more than a year ago, had ever been seen. Again, let us look at the practical results of the 40,000 salmon eggs which were sent to me by the courtesy of Professsr Baird in March, 1872.

These eggs were received by me at my hatchery, then at Clarkston, in good condition, and so far advanced that they hatched after lying in ice-water for some twenty days. It was, however, the fact of a change in the weather by which the temperature was carried up to 55°, that hastened the hatching, bringing the fish out about April 10th. Is it not a question of weighty importance whether the eggs of all the salmonoid family, may not be sufficiently retarded in their incubation, by the use of ice, as to bring the fish into life at such a time as their natural food shall he furnished through natural causes?

In proof of this theory being practical, allow me to refer you to the experiment made by me last winter at my hatchery at Clarkston, of which our State Fish Commissioners and many other distinguished citizens were cognizant.

After keeping several thousand of the white fish ova about two weeks in ice-water, I removed them to a refrigerator, so arranged as to admit alternate layers, first of moss, and then thin muslin cloth, next the ova, thin muslin cloth again, which was followed by a layer of ice, supported by galvanized iron.

This arrangement can be repeated *ad infinitum* By this means the eggs were kept at a temperature just above freezing, which delayed their incubation to the time I desired to remove them to the hatching-trough.

When hatched out they must, of course, lie in the natural element to escape death.

Some time in the month of March, these eggs were taken from the cold bath in which they had lain nearly four months in a perfectly healthy condition, and were all hatched within six days after. Their conditions were thus changed; this being true, I would ask, why may not the salmon ova be taken from the McCloud river, California, and be similarly treated, and removed to almost any part of the globe, successfully?

But to return to the fish of which we were speaking. As soon as the umbilical sac was absorbed, I deposited a large portion of them in Au Sable river—a beautiful, clear and rapid stream of water which empties into Lake Huron, near Thunder Bay. This was about May 20th—temperature of water 60 degrees.

These salmon have been caught and samples of them sent, by D. H. Fitzhugh, Esq., of Bay City, to Prof. Baird, who pronounced them genuine Penobscot salmon. These two opposite results, as relating to the Atlantic and Pacific salmon, indicate plainly that the former have fallen into waters congenial to them, while the latter, it is to be feared, have perished through the unnatural circumstances they have been forced to occupy.

In conclusion, I call especial attention of scientific and practical fish-breeders to the ideas herein suggested. For my part, I have been guided to the views I have endeavored to present, by the experiments of a humble seeker after true and rational methods of advancing what I can but feel is already and destined to become a more and more important branch of industry in this country. Hoping that what I have presented will tend to awaken a more general interest among you on this subject, and thanking you for the opportunity you have accorded me of addressing you, I close.

www.ingramcontent.com/pod-product-compliance
Lightning Source LLC
LaVergne TN
LVHW020634110826
845149LV00004B/1177
9781418191573